Copyright © 2024 Ryan Nicholson

This is a work of fiction and a product of the author's imagination. Any resemblance to real people, historical events, or places, is coincidental.
All rights reserved. No part of this book text may be reproduced in any form without permission from the author, except as permitted by U.S. copyright law.

To request permission, contact Rmnicholson@connorneptune.com
Written by Ryan Nicholson
Cover art and illustrations created through Midjourney Edited by Ryan Nicholson

Library of Congress Registration Number TXu 2-436-325

Kindle Direct Publishing

ISBN: 9798337721705   (Paperback)

Connorneptune.com

Connor Neptune was back at home, and he started to feel the need again to roam. Where to roam to he wasn't sure, but he knew going back into space was the only cure.

He had visited Jupiter and saw its wonder. Neptune was cold and distant, but it was too soon for that far an adventure.

Deep space would have to wait, for he didn't want to travel that far. Connor knew in that instance, that he would travel to Mars.

The red planet with rocks just like back home, so Connor made up his mind to Mars he would roam.
He closed his eyes and thought of the place,
Where he wanted to travel in outer space.

There it was in front of his face, the little red planet that was floating through space. Unlike the other two planets he visited before, Mars had a rocky surface for him to explore.

Connor went in for a closer view and was the first to put a footprint on the surface with his shoe. Earth was a small dot to the naked eye, and the Sun seemed half the size when seen in the Martian sky.

Red rocks covered the ground, while dust and sand were all that could be found. But the surface wasn't flat at all, instead Mars was covered with mountains both small and tall.

The mountains were so tall they didn't seem real, and they made the mountains on Earth look like a small hill.
The tallest one of all was a volcano called Olympus Mons. It towered over the tallest mountain on Earth by two and a half times.

Looking to the top of the mountain and into the sky, Connor noticed the two moons of Mars floating by. Phobos and Deimos are what the two moons were called, and of all the moons Connor had seen, these two were the strangest of all.

Unlike the Earth's Moon,
they weren't perfectly round.
Instead, they were shaped like
two big rocks that one would
find on the ground

Connor had walked for many miles to explore the land. He saw the tallest mountains and deepest canyons. As he walked, Connor heard a sound echoing through the hills. It sounded robotic as if it had wheels.

Upon further exploring he saw a robotic machine. It was sent by scientists from Earth to collect what people had never seen.

It collected samples from the ground and took pictures of the scenery. As he watched it at work, it made Connor miss his friends and his family.

When he went to Jupiter, he left filled with wonder and awe.

Then he visited Neptune and saw a cold, harsh world where no one could live at all.

Mars was different than the other two planets where he'd been.
In theory, people could live on Mars based on what Connor had seen.
And the little robot collecting rocks was proof that some people on Earth thought the same thing.

His sisters came running into his room with a smile on their face. "Where have you been Connor? We looked all over the place!"

With excitement in his voice, he told of his adventures. Connor knew his next one should be with his sisters.

This book is dedicated to my children. Adventures are always better when shared with the people in your life who love you the most!

Connor learns more about each planet he travels to in addition to learning a little more about life.

What was the lesson you learned from reading the book? We would love to hear your thoughts and what you learned.

With your parents or a grownups help, tell us what you learned at Connorneptune.com.

Don't forget to join Connor's adventures in other books in the series.

Until the next adventure, Happy Reading!!!

www.ingramcontent.com/pod-product-compliance
Lightning Source LLC
Chambersburg PA
CBHW051825210526
45473CB00005B/1744